The
Good AND THE Beautiful

THE GOOD AND THE BEAUTIFUL

FLOWER
Study

Written by
Maggie Felsch
& Molly Sanchez

TABLE OF CONTENTS

INTRODUCTION

Have you ever walked past a lovely flower and wished you knew its name? Does your child "stop and smell the roses" and truly marvel at the wonder of flowers—or just pass them by? Whether you and your child are plant experts or just beginning your journey into the splendor of flowers, this book is for you! Not only does *The Good and the Beautiful Flower Study* teach you and your child the names and features of 40 flowers, it also **encourages you to look closely** at the magnificent design of each bloom, **engages your minds**, **captures your imaginations**, and **creates a sense of wonder** for God's spectacular creations.

Use this book to discover the names and details of flowers in your neighborhood; flip through it to choose flowers to plant in your yard; or simply enjoy it inside on a rainy day.

Black-eyed Susan

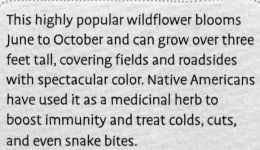

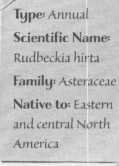

Type: Annual

Scientific Name: Rudbeckia hirta

Family: Asteraceae

Native to: Eastern and central North America

This highly popular wildflower blooms June to October and can grow over three feet tall, covering fields and roadsides with spectacular color. Native Americans have used it as a medicinal herb to boost immunity and treat colds, cuts, and even snake bites.

⊹ Flower Study ⊹

Black-eyed Susans look like genuinely curious flowers; the golden petals seem to be holding the eye up to get a prime view of its surroundings. The dark, cone-shaped center is framed by happy rays of rich sun. Count the petals of the flowers to the left. Do all Black-eyed Susans have the same number of petals?

Bluebell

Thriving in shady areas, such as the forest floor, the dazzling blue flowers provide a charming ground cover. In early spring they start out a rich cobalt blue but later turn a light powdery blue, heralding the coming of summer. They have a strong, sweet scent. There are several unrelated flowers all with the common name of bluebell, which is why the scientific name is so important.

➤ Flower Study ◄

Visualize walking down the path in the picture, the ground carpeted in perfumed blue. Bend down and let one tiny bell rest on your finger. See how delicately its petals curl up at the ends, like a ribbon you curled with scissors.

Type: Perennial

Scientific Name: Hyacinthoides non-scripta

Family: Asparagaceae

Native to: Europe

California Poppy

Nicknamed "Pot of Gold," the California state flower loves warm sunshine, so don't expect them to come out to play on cloudy days! At nighttime they tuck themselves into bed. Their four petals close up in a spiral and stay tightly curled around each other until the warm sun gently entreats them to open up again.

▸ Flower Study ◂

Notice how the poppy sits on top of a disc-shaped receptacle. It looks like a platter or pedestal with which the stem holds the papery blossom up to the sun. The fern-like, blue-green leaves contrast beautifully with each vibrant blossom.

Type: Perennial or Annual
Scientific Name: Eschscholzia californica
Family: Papaveraceae
Native to: United States and Mexico

Carnation

Christian legend tells that carnations sprang from the ground where Mary's tears fell as she watched her Son suffer. Pink carnations have long been a symbol used to celebrate motherhood. The scarlet carnation is the state flower of Ohio, in honor of William McKinley, who was a governor of Ohio and also a US president. He often wore a scarlet carnation on his lapel.

Type: Perennial
Scientific Name: Dianthus caryophyllus
Family: Caryophyllaceae
Native to: Mediterranean region

⟩ Flower Study ⟨

The carnation looks like the underside of a ball gown with layers and layers of lace. It sits atop a sturdy, gray-green stem that is so strong it almost seems to be lined with wire. Try placing a white carnation in water with food coloring. The stem will act like a straw, slurping up the color and spreading it through the lacy petals.

Type: Perennial

Scientific Name: Chrysanthemum spp.

Family: Asteraceae

Native to: East Asia and Northeastern Europe

Chrysanthemum

These popular flowers nicknamed "mums" come in many colors and color combinations. They can be daisy-like or appear more like pom-poms. A natural insecticide, which is not harmful to mammals, is made from an extract taken from chrysanthemums.

» Flower Study «

Chrysanthemums are mini fireworks on the ground! See how the petals explode out from the center, boasting their bright colors as they burst out in every direction?

Columbine

A favorite among hummingbirds, columbines are hardy wildflowers found all over the northern hemisphere.

⟫ Flower Study ⟪

Columbines come in a variety of lovely colors and are quite easy to recognize once you're familiar with their unique shape. Look at the picture below. Every columbine has five modified petals with both a blade and a spur. The petal spur is shaped like a long horn ending with a knob and is filled with nectar, perfect for hummingbirds and long-tongued insects. The petals are surrounded by five colorful sepals in the shape of a star.

Type: Perennial
Scientific name: Aquilegia spp.
Family: Ranunculaceae
Native to: Northern hemisphere

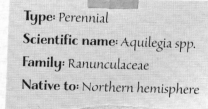

Petal Spur ⋯⋯○

Petal Blade ⋯⋯○

Sepal ⋯⋯○

7

Coneflower

Type: Perennial

Scientific Name:
Echinacea spp.

Family: Asteraceae

Native to: Eastern
and central North
America

Also widely known by its scientific name, the coneflower is used by many as a medicinal herb. This cheery flower is popular not only with people but also with pollinating insects. Purple is the most common color of echinacea, but a few other colors have been cultivated.

▸ Flower Study ◂

Each sturdy stem of the coneflower plant has a single flower with 15 to 20 drooping pinkish-purple petals surrounding the spiky central "cone." It's sometimes hard to see, but each petal has three tiny notched "teeth" at the tip. Can you see that on any petals in the picture to the left?

Cornflower

Thriving in the same conditions as corn—open sunny fields with moderately dry soil—cornflower plants used to cover corn fields in Europe with their bright and cheery blue heads. Modern use of herbicides has drastically reduced their numbers.

⇒ Flower Study ⇐

This small, electric-blue wildflower is bursting with ragged tissue paper-like petals. Make a circle with one hand by touching the tip of your first finger to the tip of your thumb. That's how big this flower head is! This little cutie is also commonly called a bachelor's button.

Type: *Annual*

Scientific Name:
Centaurea cyanus

Family: *Asteraceae*

Native to: *Europe*

Type: Perennial
Scientific Name: *Crocus spp.*
Family: Iridaceae
Native to: North Africa, the Middle East, and Southern Europe

Crocus

In areas where snow covers the ground in winter, the brave little crocus flower is usually the first to peek up out of the earth in the spring, frequently with snow still blanketing it. The kitchen spice saffron comes from the female reproductive organs of the crocus.

❧ Flower Study ❧

How many petals do you count on each crocus flower? Did you count six? The petals form a cup-shaped flower surrounded by little leaves that look much like grass blades. Crocuses are usually yellow, white, or purple and only grow about four to six inches tall.

Type: Perennial

Scientific Name: Narcissus spp.

Family: Amaryllidaceae

Native to: Northern Europe

Daffodil

Daffodils grow from bulbs that should be planted in the fall because the cold of winter promotes root development. The leaves and bulbs have poisonous crystals in them, so animals and bugs leave the flower alone. That fact and its hardy beauty make the daffodil a favorite among gardeners.

▸ Flower Study ◂

The corona is the trumpet in the center that calls out "Look at me! Look at me!" You can always count six perianth (petals) on a daffodil, opening wide their arms, as if to embrace the onlooker with a warm hug. Just imagine a six-armed hug from a lovely flower!

Perianth ·········○

Corona ·········○

11

Dahlia

Did you know that the dahlia was originally classified as a vegetable, not as a flower? That's because the root of the flower is an edible tuber, kind of like a potato! The undeniable beauty of the flower top, though, has made them popular in gardens worldwide. Dahlias come in almost any color and size; some are as small as a button and some can grow as big as your head!

❧ Flower Study ❧

Look closely at the petals of the dahlia. Do you see how each petal curls into itself? Notice how the innermost petals curl the most. The many rows of petals on the dahlia flower create a delightful pom-pom look.

Type: Perennial

Scientific Name: Dahlia spp.

Family: Asteraceae

Native to: Mexico and Central America

Dandelion

Each part of the dandelion—flower, root, and leaf—is useful to humans as food, medicine, or dye. Not only are they great for us, they also produce pollen and nectar early in the spring for honeybees, which then pollinate our plants and make delicious honey.

❧ Flower Study ☙

Dandelions are easy to identify with their sunbursts of countless little yellow petals on tubular stems. Next time you see a dandelion flower, rub it on the back of your hand and see what happens. Next time you see a dandelion seed head, blow the seeds and watch each little umbrella-like seed float away.

Type: Perennial

Scientific Name: Taraxacum spp.

Family: Asteraceae

Native to: Europe and Asia

Forget-me-not

These tiny blue flowers with yellow centers burst from fuzzy stems throughout summer and fall. In the wild they can be found near brooks, streams, and rivers. As the name implies, they are a symbol of remembrance.

➤ Flower Study ◄

1. Look at picture 1 below. When you look very closely at the tiny forget-me-not flower, you will see what looks like a tiny yellow flower in the center, surrounded by a white star. Isn't it beautiful? 2. Look at the whole flower. How many petals does it have? 3. Look how pretty the cluster of flowers is together. What color are the flower buds before opening up? 4. Whenever you see these lovely flowers, don't forget to pause and appreciate their beauty!

Type: Usually biennial, depending on the climate
Scientific Name: Myosotis spp.
Family: Boraginaceae
Native to: Temperate Europe, Asia, and North America

Gardenia

Gardenia, which grows in tropical and subtropical climates, is part of the coffee family. The creamy-white flowers of these woody evergreen shrubs have an intensely wonderful fragrance.

❧ Flower Study ❧

Look at the white gardenia petals in the picture above. Imagine plucking one of the petals off and rubbing it gently across your cheek. Can you imagine how velvety soft it must feel? Notice the dazzling spiral shape of every gardenia flower. Even the flower buds are in the form of a spiral!

Flower Buds

Geranium

Commonly grown in flower pots and garden beds, Geraniums provide low-maintenance blooms with full foliage, and they are tolerant of heat. Their vast assortment of colors can provide a kaleidoscope of shades for window boxes and balconies.

» Flower Study «

Bold balls of flowers on the end of long stems are reminiscent of sweet, syrupy snow cones on a hot summer day. The flowers don't smell very strong, but their cheerful balls of bright colors make us want to lean in for a closer look anyway!

Type: Annual
Scientific Name:
Pelargonium x hortorum
Family: *Geraniaceae*
Native to: *South Africa*

Hibiscus

This delicate tropical blossom is traditionally worn by girls on the Hawaiian and Tahitian Islands. The large, bold flower grows up to six inches across. Many cultures use the dried flower and seed pods to make a tart tea, full of medicinal value, rich in color, and high in vitamin C.

▸ Flower Study ◂

Look at the pistil that bursts from the center of the hibiscus flower. Doesn't it look like a satellite tower? Can you imagine climbing it? Pretend you are a bee, butterfly, hummingbird, or bat, eager to enjoy its sweet nectar.

Type: *Annual or Perennial*

Scientific Name: *Hibiscus rosa-sinensis*

Family: *Malvaceae*

Native to: *Tropical and subtropical regions*

Type: Perennial
Scientific Name: Hyacinthus spp.
Family: Asparagaceae
Native to: Mediterranean region

Hyacinth

Hyacinths are some of the easiest flowers to grow, and they give a lot of bang for your buck with their clusters of blooms, as well as strong fragrance. As with daffodils, bugs and rodents stay away, letting them boast their beauty all season. They do attract butterflies!

Flower Study

Don't hyacinths just seem like the friendliest flower? They don't mind being crowded together on the same spike, called a raceme. "Hi! Hi! Hyacinth!" they seem to call out to each other as they reach around to hold hands.

Hydrangea

Hydrangeas have a very unique feature: the flowers change colors based on how acidic or alkaline the soil is! Acidic soil causes the flowers to be blue, and alkaline soil yields pink flowers. When the soil is in-between, the flowers are purple!

❧ Flower Study ❧

Look at picture 1. Imagine walking down this lovely path surrounded by hydrangeas. What do you imagine it would look like, feel like, and smell like? 2. Now look at the shape of an individual "pom-pom" formed by a perfect cluster of flowers. 3. Hydrangea flowers have four to five paper-thin petals. Notice the beautiful center, too. 4. When you look very closely, the center looks finer than a queen's jewels.

Type: Perennial
Scientific Name: Hydrangea spp.
Family: Hydrangeaceae
Native to: Asia and America

Indian Paintbrush

The petals of this common wildflower are edible and were eaten regularly by many Native American tribes. The plant was also made into a hair wash.

⟩ Flower Study ⟨

Imagine sitting in the cool grass of a mountain meadow or the soft sand of a calm desert. Now imagine reaching out and plucking a paintbrush from the ground to paint the magnificent scene. Though the Indian paintbrush doesn't really paint, the wildflower has captured the imagination of countless dreamers because of the brush-like flowers that seem to be dipped in red paint.

Type: Annual and Perennial
Scientific Name: Castilleja spp.
Family: Orobanchaceae
Native to: North America and Northern Asia

Iris

The showy iris flower comes in a vast rainbow of colors, which is why it got its name from the Greek word for rainbow. Bearded irises, like those to the right, are popular among gardeners.

⟩ Flower Study ⟨

An iris looks a bit like a three-armed Spanish dancer wearing a ruffled gown, complete with full, frilly sleeves thrust up in the air. If you want to, pretend you are a Spanish dancer with a full skirt that flounces out as you spin.

Type: Perennial
Scientific Name: Iris spp.
Family: Iridaceae
Native to: Northern hemisphere

21

Lavender

The beautifully flowering lavender plant is part of the mint family. It is widely known for its sweet floral aroma, undeniable beauty, and many uses as a medicinal herb. It is commonly used to ease anxiety and promote healthy sleep.

❧ Flower Study ❧

Mounds of long stems covered in clusters of tiny purple blooms and its unforgettable fragrance are the distinguishing features of lavender. There are other kinds of purple flowers clustered on long stems, so look closely at the shape of the flowers and the long slender leaves.

Type: Perennial

Scientific Name: Lavandula spp.

Family: Lamiaceae

Native to: Mediterranean region and Middle East

Lilac

Lilacs grow on bushes that can grow as tall and as wide as a small house! Think carefully where you want to plant a lilac bush, because it can live 200 years! The summer blossoms smell incredible. Try never to miss a chance to sniff blooming lilacs if you pass them.

Type: Perennial

Scientific Name: *Syringa spp.*

Family: *Oleaceae*

Native to: Eastern Europe and Asia

❧ Flower Study ❧

Grab a piece of paper and a pencil and draw a lilac flower. Look closely—is your drawing perfectly symmetrical? Take a moment to consider how amazing it is that each of the tens of thousands of simple four-petal blossoms is perfectly symmetrical!

Lily

Lilies are treasured garden flowers with a marvelous display. Single sturdy stems grow from bulbs and burst with blossoms in the summer. Offering dreamy scents and large petals, lilies are also splendidly popular as cut flowers. They flourish in the sun.

❧ Flower Study ❧

Think in threes with a lily. It has three petals surrounded by three sepals (they look just like the petals). It has three plus three stamens (they hold the golden pollen), and the stigma in the center has three points. Can you see all the threes?

Type: Perennial
Scientific Name: Lilium spp.
Family: Liliaceae
Native to: Northern hemisphere

Lily of the Valley

These dainty, fragrant, bell-shaped flowers become red berries after the petals drop. But don't eat them! Every part of this plant is poisonous. It makes a charming ground cover for cooler, shaded areas.

❧ Flower Study ❧

If you were the size of a field mouse, wouldn't it be lovely to pretend you were in a symphony playing these bells? Picture yourself sitting on the strappy green leaf. Imagine that each delightful bell has a different sound, and you create a melody as lovely to the ear as the flower is to your sight and smell.

Type: Perennial

Scientific Name: *Convallaria majalis*

Family: *Asparagaceae*

Native to: Asia and Europe

Lupine

Lupines are a type of legume (bean). Some varieties grow pods of beans called lupini beans. When prepared properly, these highly nutritious beans are packed with protein and fiber and are valued as a superfood in some countries. Lupines grow wild freely across North America.

One of the most distinctive parts of the lupine is its leaves. Notice how they look like an open hand with many fingers. The unique flower resembles a praying bird, with its beak reaching heavenward and the wings pressed together like praying hands. Do you see it?

Type: Perennial or Annual
Scientific Name: Lupinus spp.
Family: Fabaceae
Native to: North America

Marigold

Not only do marigolds brighten a garden with their cheery pops of orange-gold, but they also provide many benefits to the garden, including repelling unwanted insects. The flowers are edible and have herbal medicinal uses as well.

✿ Flower Study ✿

Imagine cupping a marigold in the palm of your hand. In your mind, run your finger through the velvet-soft petals. See how wavy they are all around the edges? If you wanted to, you could pluck a petal and taste it. Depending on the variety, it might taste spicy like pepper or citrusy like an orange.

Type: *Annual or Perennial*
Scientific Name: *Tagetes spp.*
Family: *Asteraceae*
Native to: *Mexico*

Orchid

There are 28,000 species of orchids. That's more than twice the number of bird species and four times the number of mammal species! Orchid plants range in size from two millimeters to the largest in the world recorded at 7.6 meters tall! Vanilla beans come from one species of orchid.

⟩ Flower Study ⟨

One orchid petal, the labellum, is modified and usually enlarged. It acts as a pedestal for pollinators. Look at the labellum in the picture below. Doesn't it look like a bee? Its purpose is to attract bees!

Type: Perennial
Scientific Name: Orchidaceae spp.
Family: Orchidaceae
Native to: Tropical forests

Pansy

Pansies are the most popular edible flower, eaten in fresh salads or candied in desserts. Lest you think they are wimpy, perhaps implied by their name, pansies are wonderfully hardy in the cold and can withstand light freezes and snow. Next time you buy lettuce for a salad, see if you can get some pansies to add to it! Or grow both in a garden!

≽ Flower Study ≼

Once you know a pansy, you will always be able to recognize it by its unique petal configuration and its sunburst at the center. How many colors of pansies can you count just on this page?

Type: Perennial
Scientific Name: Viola tricolor
Family: Violaceae
Native to: Europe and Asia

Peony

Pronounced PEE-uh-nee, these delightful flowers have graced gardens for centuries. In fact, there are written records from eighth century China describing the dependable and enchanting peony.

❧ Flower Study ❧

The dense layers of thickly lush petals make the soft, round face of a peony unforgettable. Can you imagine closing your eyes and pressing your face into these gloriously soft and fragrant petals? How do you think it would feel?

Type: Perennial
Scientific Name: Paeonia spp.
Family: Paeoniaceae
Native to: Europe and Asia

Petunia

Petunias are one of the most popular garden flowers. They make colorful borders or spill over containers and hanging baskets. They bloom abundantly all spring and summer, except in extreme heat. Some varieties require deadheading (pulling off dead flowers) to keep them blooming. You can find them in nearly every color!

◆ Flower Study ◆

Petunias kind of look like a bat starfish, or webbed starfish. Can you see the star in the flower? The green stems and leaves can be sticky. Imagine pinching off the dead blossoms and getting sticky fingers as you touch the fragrant, colorful plant.

Type: *Annual*
Scientific Name: *Petunia spp.*
Family: *Solanaceae*
Native to: *South America*

Plumeria

You may recognize these flowers as those sometimes used to make Hawaiian leis because of their large size, lovely scent, and appealing colors. They are most fragrant at night to entice sphinx moths to pollinate them. They are tricking the moths; they actually have no nectar.

Type: Perennial
Scientific Name: *Plumeria spp.*
Family: *Apocynaceae*
Native to: Central America, Mexico, and the Caribbean

⟩ Flower Study ⟨

Have you ever held a plumeria? They have a sturdy, rubbery feel. Look at the center of the flower and how it extends outward in a fantastic spiral. Imagine stringing plumerias together to make a beautiful lei. What colors would you use? To whom would you give your fragrant lei?

Type: Perennial or annual
Scientific Name: Ranunculus spp.
Family: Ranunculaceae
Native to: North America

Ranunculus

If you want to grow flowers perfect for cutting and using in arrangements, try the ranunculus. Its tall stem is perfectly straight and strong, and the mesmerizingly beautiful flower heads, which come in a variety of bold colors, are enough to make any bouquet glow.

⊱ Flower Study ⊰

Can you imagine trying to count the layers and layers of soft, thin ranunculus petals, all tightly bound together into one perfect flower head? How many petals do you think one flower can hold?

Type: Perennial
Scientific Name: Rosa spp.
Family: Rosaceae
Native to: Mostly Asia

Rose

The red rose is the universal symbol of love around the world. Different colors have different meanings when given as gifts. Roses are usually grown ornamentally, and most are very fragrant. Though roses are usually nontoxic, they have a stronger defense system: spiny thorns!

⟩ Flower Study ⟨

If someone hands you a rose, what is the first thing you want to do? Smell it! Can you remember the strong, sweet fragrance of a rose? Imagine brushing your nose gently against the soft petals that overlap one another in a swirl of intricate beauty.

Snapdragon

These hardy flowers grow wild across rocky areas of North America, Asia, Europe, and Northern Africa. The flowers bloom starting from the bottom of the stem to the top. Once they have all bloomed, they make a magnificent display! Snapdragons come in just about every color imaginable.

⟩ Flower Study ⟨

Snapdragons are named for their resemblance to the mythical creature. Next time you come across a snapdragon, reach out and pinch the flower gently on each side with your thumb and forefinger; the "mouth" will open with each squeeze!

Type: Annual or Perennial

Scientific Name: Antirrhinum spp.

Family: Plantaginaceae

Native to: Europe, North Africa, and North America

Sunflower

Is it any wonder how these charming, bright flowers got their name? Hardy enough to thrive in dry, rocky soil, sunflowers brighten a variety of landscapes from mountains to deserts and from abandoned fields to roadsides. Once the seeds ripen, they become a cherished treat for birds and people alike.

⋗ Flower Study ⋖

It's natural to admire the bright yellow petals of a sunflower. But look closely at the center of the sunflower head; let your eyes follow the phenomenal geometrical pattern of the seeds. Isn't it amazing? How difficult do you think it would be to create that pattern yourself by sticking seeds into clay?

Type: *Annual*

Scientific Name: *Helianthus annuus*

Family: *Asteraceae*

Native to: *North and Central America*

Sweet Pea

Sweet peas are a climbing plant with curling tendrils that grab onto any supportive structure. The plant can grow to a height of three to six feet. They are usually grown from seeds, which are the size of a pea. They are known for their strong, sweet perfume.

⟩ Flower Study ⟨

Sweet peas look like tiny triceratopses, without the horns. The top petals resemble the bony frill (like a fan) of the dinosaur, and the tiny petals in front, the beak. Can you imagine colorful, itty bitty triceratopses climbing the side of your house?

Type: Annual

Scientific Name: *Lathyrus odoratus*

Family: Fabaceae

Native to: Mediterranean region

Tulip

Tulips grow from bulbs. You've probably eaten a different kind of bulb: an onion! Like many other flowers, tulips are edible!

Flower Study

Look at the picture on the right. Every tulip has three petals and three sepals. On most flowers the sepal is green and surrounds the bottom of the flower. Tulip sepals, however, look almost identical to tulip petals in color and size, so tulips look like they have six petals. Another recognizable feature of the tulip is its long, straight, tubular stem and long, wide leaves. Tulips come in almost every color imaginable.

Type: Perennial

Scientific Name: Tulipa spp.

Family: Liliaceae

Native to: Central and Eastern Asia

Sepal · · · · · Petal

Verbena

Itty-bitty, pinkish-purple flowers bunch together on the tops of semiwoody stems and send an exhilarating fragrance out into the gentle breeze. They beckon pollinators larger than themselves to come and gather nectar. Verbena grows wild, even in dry conditions.

▶ Flower Study ◀

Look at the picture with the bee. Verbena flowers are just the right size for a bee to pick a bouquet for a friend if she could. How many petals do you count on the verbena? Did you count five? Each petal is shaped like a little heart, perfect for Valentine's Day.

Type: Annual and Perennial
Scientific Name: Verbena spp.
Family: Verbenaceae
Native to: Asia and the Americas

Violet

The common blue violet grows wild throughout the eastern part of North America, particularly in shaded woods, thickets, and along stream banks. It is the state flower of three states: Rhode Island, Illinois, and New Jersey.

Flower Study

Picture yourself walking down a shaded forest path surrounded by a dense floor of violet flowers. Now bend down and look closely (look at the picture to the right). Do you see how the bottom petal is the perfect shape to place your thumb? It's also the perfect place for a honeybee to land and gather nectar!

Type: Mostly perennial
Scientific Name: Viola spp.
Family: Violaceae
Native to: North America

Zinnia

Gushing with rich color, zinnias are popular garden flowers because of their wide range of colors and shapes. They are tolerant of hot summer temperatures and are easy to grow from seeds.

❧ Flower Study ❧

Pretend you are a butterfly heading to a garden full of brilliantly colored zinnias. Which color would you land on? Which do you think has the sweetest nectar? How many petals can you count on the big red zinnia pictured here?

Type: *Annual*
Scientific Name: *Zinnia spp.*
Family: *Asteraceae*
Native to: *The Americas*

41

GLOSSARY

There are three different "types" of flowers based on their life cycles. These are listed on a small "fact sheet" for each flower.

Perennial: A perennial plant lives longer than two years. Perennials include most trees, shrubs, grasses, and many flowers. The term also includes plants like geraniums (page 16), which are perennials in warm climates but are killed by frost each year. These are sometimes called "tender perennials."

Annual: An annual plant grows from seed each spring and completes its entire life cycle in a single growing season.

Biennial: A biennial plant has a two-year life cycle. It sprouts and grows the first year its seed is planted; then it produces flowers, makes seeds, and dies in its second year.

The fact sheets also include some other terms.

Scientific Name: This is the universal name used by scientists and many gardeners. Most scientific names are Latin based. Using scientific names helps clarify between species and varieties. It also helps share information between different languages.

spp: This means there are multiple species of the flower, and no particular species is specified in this book.

Family: A flower family is a collection of flowers that share certain characteristics and are grouped together through classification. Flowers can be categorized by shape, seed groupings, root formation, and more.

Native to: This is where the plant originated, though most have spread around the world because of modern transportation and global trade.

INDEX BY FAMILY

Printed in Canada